ollins

The Shanghai Maths ct

For the English National Curriculum

一课一练

Year 4 Learning

Collins

William Collins' dream of knowledge for all began with the publication of his first book in 1819.

A self-educated mill worker, he not only enriched millions of lives, but also founded a flourishing publishing house. Today, staying true to this spirit, Collins books are packed with inspiration, innovation and practical expertise. They place you at the centre of a world of possibility and give you exactly what you need to explore it.

Collins. Freedom to teach.

Published by Collins
An imprint of HarperCollins*Publishers*
The News Building
1 London Bridge Street
London
SE1 9GF

Browse the complete Collins catalogue at
www.collins.co.uk

© HarperCollins*Publishers* Limited 2018

10 9 8 7 6 5 4 3 2 1

978-0-00-822598-8

The authors assert their moral rights to be identified as the authors of this work.

Learning Books Series Editor: Amanda Simpson

Practice Books Series Editor: Professor Lianghuo Fan

Authors: Laura Clarke, Caroline Clissold, Sarah Eaton, Linda Glithro, Paul Hodge, Jane Jones, Steph King, Richard Perring, Paul Wrangles

British Library Cataloguing in Publication Data

A catalogue record for this publication is available from the British Library.

Publishing Manager: Fiona McGlade
In-house Editor: Mike Appleton
In-house Editorial Assistant: August Stevens
Project Manager: Emily Hooton
Copy Editor: Tanya Solomons
Proofreaders: Tony Clappison and Steven Matchett
Cover design: Kevin Robbins and East China Normal University Press Ltd
Cover artwork: Daniela Geremia
Internal design: Amparo Barrera
Typesetting: Ken Vail Graphic Design and 2Hoots Publishing Services Ltd
Illustrations: QBS
Production: Sarah Burke

Printed and bound by Grafica Veneta, S.p.A., Italy

Photo acknowledgements
The publishers wish to thank the following for permission to reproduce photographs. Every effort has been made to trace copyright holders and to obtain their permission for the use of copyright materials. The publishers will gladly receive any information enabling them to rectify any error or omission at the first opportunity.

(t = top, c = centre, b = bottom, r = right, l = left)
p6 t TrotzOlga/Shutterstock, p6 tc Nadezda/Shutterstock, p6 b gresei/Shutterstock, p28 bl wavebreakmedia/Shutterstock, p28 tr Hafiz Johari/Shutterstock, p29 tr Alison Hancock/Shutterstock, p29 bl Gabrielle Ewart/Shutterstock, p36 l Aleksandr Bryliaev/Shutterstock, p40 tc NokHoOkNoi/Shutterstock, p40 br BlueRingMedia/Shutterstock, p40 tl alexmstudio/Shutterstock, p41 ltc Viktoriya Yakubouskaya/Shutterstock, p43 bc Plan-B/Shutterstock, p44 cr Macrovector/Shutterstock, p44 bc Chalintra.B/Shutterstock

MIX
Paper from
responsible sources
FSC™ C007454
www.fsc.org

This book is produced from independently certified FSC paper to ensure responsible forest management.

For more information visit:
www.harpercollins.co.uk/green

3-D shapes and their properties

Cone
1 circular base
1 curved surface
1 apex

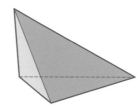

Sphere
1 curved surface

Tetrahedron
4 faces
6 edges
4 vertices

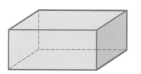

Cuboid
6 faces
12 edges
8 vertices

Cylinder
2 circular bases
1 curved surface

Cube
6 faces
12 edges
8 vertices

Triangular prism
5 faces
9 edges
6 vertices

Square-based pyramid
5 faces
8 edges
5 vertices

3-D shapes have faces, edges and vertices. The flat surfaces are called faces and these are polygons. The line where two faces meet is called an edge. A vertex is a point where three edges meet. Some 3-D shapes cannot be described by their faces, edges and vertices. One example is a cone, which has one circular base and one curved surface that connects the circular base to a pointed tip called the apex.

Converting between different units of measurement

Converting larger units to smaller units

To convert a larger unit to a smaller unit (for example, kilometres to metres), find the number of smaller units needed to make one larger unit. Then multiply that number by the number of larger units.

4 km = 4000 m (4 × 1000 = 4000)

Converting smaller units to larger units

To convert a smaller unit to a larger unit (for example mm to cm), divide by the number of smaller units needed to make one larger unit.

50 mm = 5 cm (50 ÷ 10 = 5)

The conversion values are given in the table below.

Quantity	Name of unit	Symbol	Value
length	millimetre	mm	
	centimetre	cm	10 mm = 1 cm
	metre	m	100 cm = 1 m
	kilometre	km	1000 m = 1 km
mass	gram	g	
	kilogram	kg	1000 g = 1 kg
volume	millilitre	ml	
	litre	l	1000 ml = 1 l
time	second	s	
	minute	min	60 s = 1 min
	hour	h	60 min = 1 h
	day	day	24 h = 1 day
	week	week	7 days = 1 week
	year	year	365 days = 1 year (366 days in a leap year)

Units of measurement

Comparing division calculations

When the divisor stays the same, a smaller dividend means the quotient will be smaller too.

When the dividend stays the same, an increasing divisor means the quotient decreases.

dividend	divisor	quotient
24 ÷	6 =	4

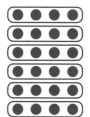

dividend	divisor	quotient
18 ÷	6 =	3

	dividend	divisor	quotient
	18 ÷	2 =	9
	18 ÷	3 =	6
	18 ÷	6 =	3
	18 ÷	9 =	2

Fractions of shapes and quantities

The orange is divided into 2 equal parts. Each part is $\frac{1}{2}$ of the orange.

The chocolate bar is divided into 12 equal parts.
Each part is $\frac{1}{12}$ of the chocolate bar.

The hexagon is divided into 6 equal parts. Each part is $\frac{1}{6}$ of the hexagon.

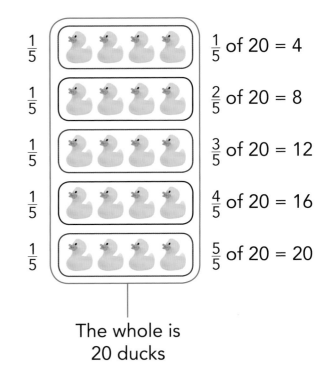

$\frac{1}{5}$ $\frac{1}{5}$ of 20 = 4

$\frac{1}{5}$ $\frac{2}{5}$ of 20 = 8

$\frac{1}{5}$ $\frac{3}{5}$ of 20 = 12

$\frac{1}{5}$ $\frac{4}{5}$ of 20 = 16

$\frac{1}{5}$ $\frac{5}{5}$ of 20 = 20

The whole is
20 ducks

Rounding numbers

Here's a number:

65 853

Here's the number rounded in three ways:

To the nearest 10, it's	To the nearest 100, it's	To the nearest 1000, it's
65 850	**65 900**	**66 000**

65 853
←————|————————→
65 850   65 860

65 853
←————————|————→
65 800   65 900

65 853
←————————————|——→
65 000   66 000

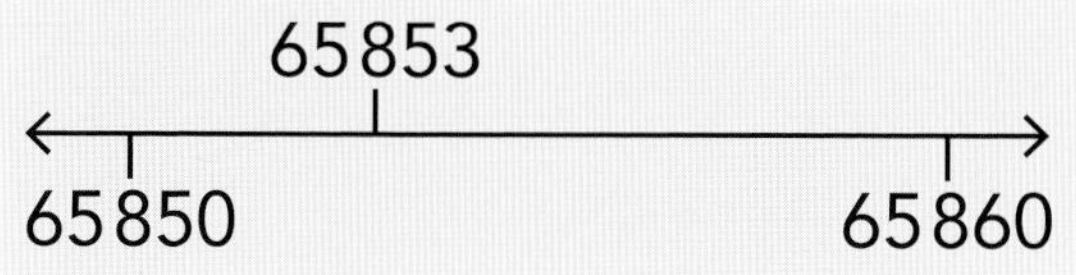

53 is closer to 50 than 60 so the number is rounded down to the nearest 10.

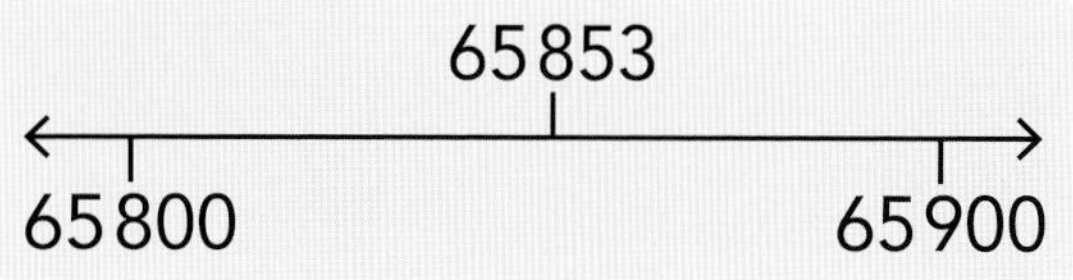

853 is closer to 900 than 800 so the number is rounded up to the nearest 100.

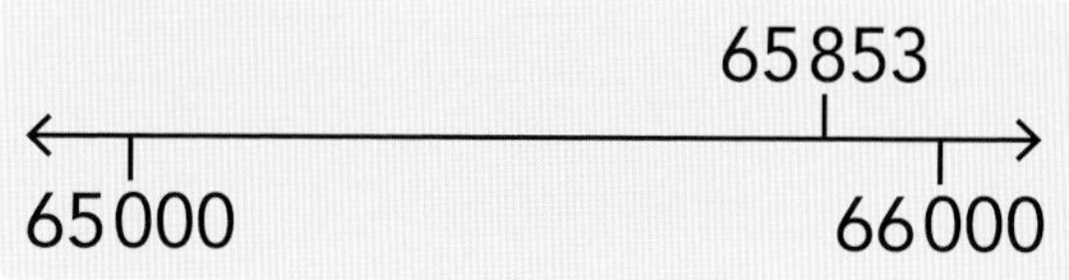

5853 is closer to 6000 than 5000 so the number is rounded up to the nearest 1000.

Don't forget!

The digit 5 is exactly in the middle, but the rule is that 5 is always rounded up.

So,

- 45 rounded to the nearest ten is 50
- 350 rounded to the nearest 100 is 400
- 7500 rounded to the nearest 1000 is 8000.

Rounding is useful … to calculate the approximate answer … to check your answer … to make describing and understanding numbers easier.

Addition and subtraction with 4-digit numbers

Example of a partitioning and recombining strategy for addition

Partition both addends, then add starting with the 1s. Find the total of the partial sums.

Example 2847 + 4536 = ☐

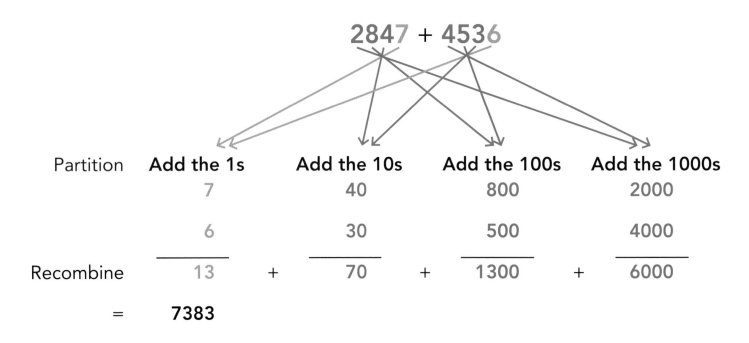

	Add the 1s	Add the 10s	Add the 100s	Add the 1000s
Partition	7	40	800	2000
	6	30	500	4000
Recombine	13 +	70 +	1300 +	6000
=	**7383**			

Example of a partitioning and recombining strategy for subtraction

Partition both numbers and subtract the 1000s, 100s, 10s and 1s in separate steps. Find the sum of the differences.

Example 7854 – 4123 = ☐

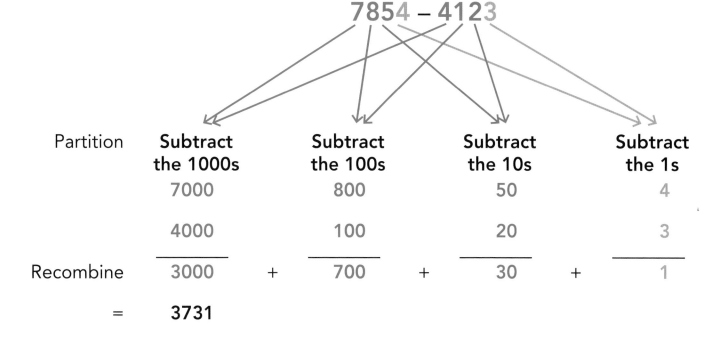

	Subtract the 1000s		Subtract the 100s		Subtract the 10s		Subtract the 1s
Partition	7000		800		50		4
	4000		100		20		3
Recombine	3000	+	700	+	30	+	1
=	**3731**						

Column addition

5439 + 1843

```
    5  4  3  9
+   1  8  4  3
_____


    5  4  3  9
+   1  8  4  3
_____
             2
          1

    5  4  3  9
+   1  8  4  3
_____
          8  2
          1

    5  4  3  9
+   1  8  4  3
_____
       2  8  2
      1     1

    5  4  3  9
+   1  8  4  3
_____
    7  2  8  2
      1     1
```

Column subtraction

5294 – 3769

```
    5  2  9  4
–   3  7  6  9
_____


                  8  1
    5  2  9̸  ⁴4
–   3  7  6  9
_____
             5

            8  1
    5  2  9̸  ⁴4
–   3  7  6  9
_____
          2  5

    4   8
    5̸ 2 9̸ ⁴4
–   3  7  6  9
_____
       5  2  5

    4   8
    5̸ 2 9̸ ⁴4
–   3  7  6  9
_____
    1  5  2  5
```

Bar models

Bar models can show the relationship between a whole and its parts.

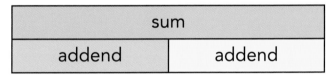

whole	
part	part

sum	
addend	addend

minuend	
subtrahend	difference

Bar models can show how different parts compare with each other.

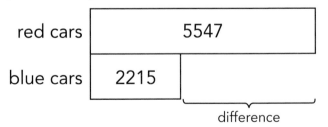

red cars | 5547

blue cars | 2215 | difference

5547 – 2215 = ☐

Mental methods

I can solve the multiplication 58 × 9 mentally in two different ways.

(1) Finding part products

9 × 58

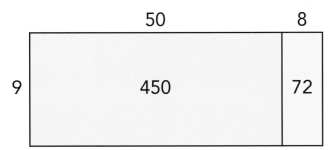

9 × 50 = 450

9 × 8 = 72

450 + 72 = 522

(2) Multiplying by a near multiple of 10

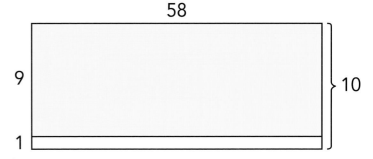

9 × 58 = (10 × 58) − (1 × 58)

= 580 − 58

= 522

Using known facts

Breaking down multiples of 10 and 100

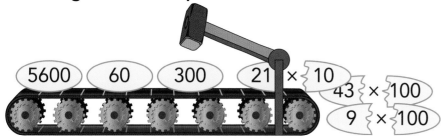

Breaking down multiples of 10 and 100 often makes a calculation easier.

$900 \times 8 = 9 \times 100 \times 8 = 9 \times 8 \times 100 = 72 \times 100$

$210 \times 4 = 21 \times 10 \times 4 = 21 \times 4 \times 10 = 84 \times 10$

$60 \times 70 = 6 \times 10 \times 7 \times 10 = 6 \times 7 \times 10 \times 10 = 42 \times 100$

Partitioning to multiply by known facts

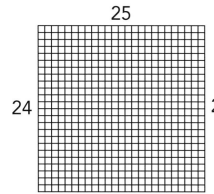

$24 \times 25 =$

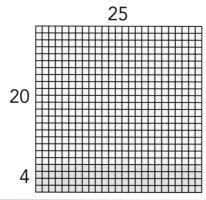

$$24 \times 25 = 20 \times 25 + 4 \times 25$$
$$= 500 + 100$$
$$= 600$$

Using factors to multiply by known facts

16 × 25 =

    = 4 × 4 × 25

    = 4 × 100

    = 400

I know that 4 × 25 equals 100

Using brackets and the order of operations

Brackets can be used to show what needs to be carried out first in a calculation.

    (30 × 6) + (4 × 6) = 180 + 24

                   = 204

Multiplication and division should be carried out before addition or subtraction.

    16 × 10 + 16 × 5 = 160 + 80

                  = 240

Making estimates

The product of 16 × 22 is greater than the product of 16 × 20 but less than the product of 16 × 25.

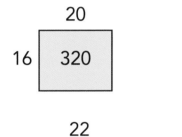

20

16 | 320

22

16 | ?

320

25

16 | 400

Using the column method

The array and the column method both represent the multiplication 6 × 42.

10	10	10	10	1	1
10	10	10	10	1	1
10	10	10	10	1	1
10	10	10	10	1	1
10	10	10	10	1	1
10	10	10	10	1	1

$$\begin{array}{r} 4\ \ 2 \\ \times\ \ \ \ 6 \\ \hline \end{array}$$

10

100 10 10 1

100 10 10 1

100 10

100

$$\begin{array}{r} 4\ \ 2 \\ \times\ \ \ \ 6 \\ \hline 2\ 5\ 2 \\ {\scriptstyle 2\ \ \ 1} \end{array}$$

Multiplying by larger numbers

```
      4   2
  ×   3   6
```

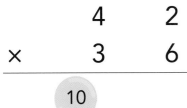

multiply 42 and 6
in the ones place

```
      4   2
  ×   3   6
```

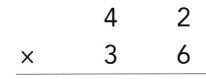
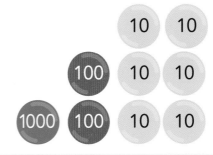

multiply 42 and 3
in the tens place

multiply 42 and 6 in the ones place
multiply 42 and 3 in the tens place

Here you can see both the steps together in one calculation.

The first answer row shows the product of 42 × 6 ones.

The second answer row shows the product of 42 × 3 tens.

Look at the use of zero to show that 2 × 3 tens is 60.

The two part products are then added to give the total product.

15

Dividing 2-digit or 3-digit numbers by tens

The arrays show 6 ÷ 3 and 60 ÷ 30

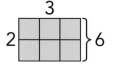 

6 ones divided by 3 ones is 2.

6 tens divided by 3 tens is 2.

60 divided by 30 is the same as 6 divided by 3.

Remainders

This shows 15 divided into groups of 6.

There are 2 groups of 6 and 3 single ones left over. So, 15 ÷ 6 = 2 r 3

Same image, different numbers, same relationship

This shows 150 divided into groups of 60.

There are 2 groups of 60 and 30 left over. So, 150 ÷ 60 = 2 r 30

Using the column method

150 ÷ 60 = 2 r 30

```
         2
60) 1 5 0

    1 2 0

  3 0
```

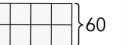

10 10 10 10 10 10   } two
10 10 10 10 10 10   } 60s

30 10 10 10

Hundredths

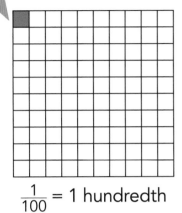

$\frac{1}{100}$ = 1 hundredth

$\frac{10}{100}$ = 10 hundredths

$\frac{1}{10}$ = 1 tenth

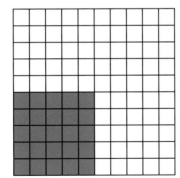

$\frac{25}{100}$ = 25 hundredths

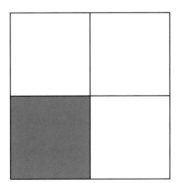

$\frac{1}{4}$ = 1 quarter

Fractions wall

1 whole is equivalent to $\frac{2}{2}$ $\frac{3}{3}$ $\frac{4}{4}$ $\frac{5}{5}$ and $\frac{6}{6}$. When the numerator and the denominator are equal, the fraction will be equivalent to 1.

$\frac{1}{2}$ is equivalent to $\frac{2}{4}$ $\frac{3}{6}$ $\frac{4}{8}$ $\frac{5}{10}$ $\frac{6}{12}$ $\frac{7}{14}$ and $\frac{8}{16}$. When the numerator is half the size of the denominator, the fraction will be equivalent to $\frac{1}{2}$.

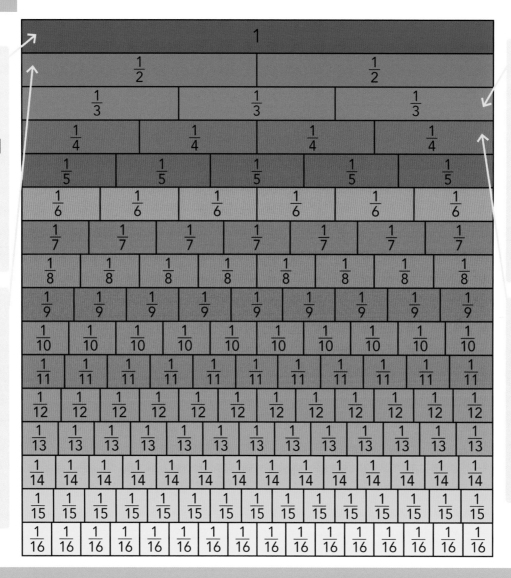

$\frac{1}{3}$ is equivalent to $\frac{2}{6}$ $\frac{3}{9}$ $\frac{4}{12}$ and $\frac{5}{15}$. When the numerator is a third of the size of the denominator, the fraction will be equivalent to $\frac{1}{3}$.

$\frac{1}{4}$ is equivalent to $\frac{2}{8}$ $\frac{3}{12}$ and $\frac{4}{16}$. When the numerator is a quarter the size of the denominator, the fraction will be equivalent to $\frac{1}{4}$.

Vocabulary

Pure decimal
(read as 'zero point seven five')

0.75

The whole number part is zero.

Mixed decimal
(read as 'seven point five zero two')

7.502

The whole number part is not zero.

Representations

Place value charts (showing 0.75)

100	10	1	.	0.1	0.01
		0	.	7	5

100	10	1	.	0.1	0.01
			.		

These place value charts show that 0.75 is made from seven 0.1s and five 0.01s.

Gattegno chart (showing 0.75)

This chart is named after the mathematician and maths teacher Caleb Gattegno who invented it. His name is said as 'Gat – ten – yo'. Here it shows that 0.75 is made from 0.7 and 0.05.

Ten Thousands	10 000	20 000	30 000	40 000	50 000	60 000	70 000	80 000	90 000
Thousands	1000	2000	3000	4000	5000	6000	7000	8000	9000
Hundreds	100	200	300	400	500	600	700	800	900
Tens	10	20	30	40	50	60	70	80	90
Units	1	2	3	4	5	6	7	8	9
Tenths	0.1	0.2	0.3	0.4	0.5	0.6	0.7	0.8	0.9
Hundredths	0.01	0.02	0.03	0.04	0.05	0.06	0.07	0.08	0.09
Thousandths	0.001	0.002	0.003	0.004	0.005	0.006	0.007	0.008	0.009

Place value counters (showing 0.75)

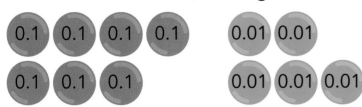

These counters show that 0.75 is made of seven 0.1s and five 0.01s

Number line (showing 0.75)

You can see here that $\frac{1}{10}$ has the same value as 0.1

0.75

| 0.1 | 0.2 | 0.3 | 0.4 | 0.5 | 0.6 | 0.7 | 0.8 | 0.9 | 1.0 |

$\frac{1}{10}$ $\frac{2}{10}$ $\frac{3}{10}$ $\frac{4}{10}$ $\frac{5}{10}$ $\frac{6}{10}$ $\frac{7}{10}$ $\frac{8}{10}$ $\frac{9}{10}$ 1

This number line shows that 0.75 is between $\frac{7}{10}$ and $\frac{8}{10}$.

In fact, 0.75 is $\frac{75}{100}$ – you can see the hundredths lines. 0.75 is on the 75th line.

A place value slider is helpful for converting between units – it shows what happens to a number when it is multiplied or divided by 10, 100 or 1000.

To multiply by 10 slide one place to the left; to divide by 10, slide one place to the right.

Drawing and reading graphs

Here is a statistical table showing the temperature on the first day of each month in Paris.

Month	Jan	Feb	Mar	Apr	May	Jun	Jul	Aug	Sep	Oct	Nov	Dec
Temp (°C)	5	6	9	11	15	16	20	20	16	12	7	5

The data has been plotted on a line graph.

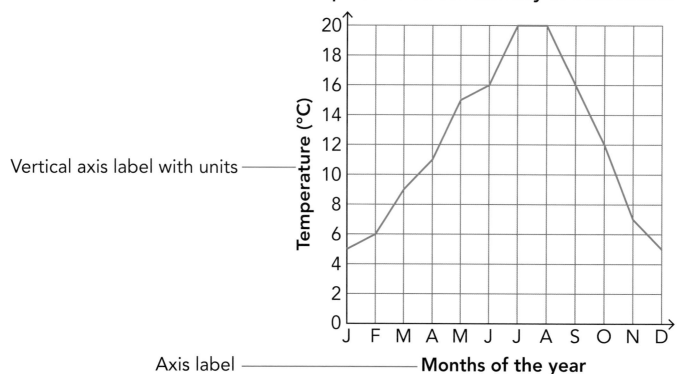

Main title ——— **Temperature on the first day of each month in Paris**

Vertical axis label with units ———

Axis label ——— **Months of the year**

Months are plotted on the horizontal axis (time is always plotted on this axis).

Temperature in degrees Celsius is plotted on the vertical axis.

The graph clearly shows temperature trends. The slope of the line is called the gradient. The steeper the gradient the more quickly the temperature is changing.

From January to July the graph shows an upward trend – the temperature is increasing each month.

The highest temperature is in July and August at 20 °C.

From August to December the graph shows a downward trend – the temperature is decreasing each month.

The gradient is steepest between October and November.

Angles

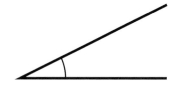

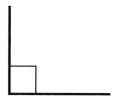

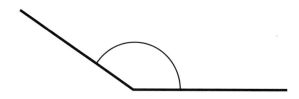

An acute angle is less than 90°.

A right angle is 90°.

An obtuse angle is greater than 90° and less than 180°.

2-D shapes

The names of 2-D shapes are usually based on the number of sides or angles.

'lateral' means number of sides

'gon' means number of angles

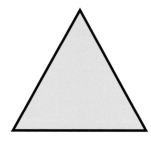

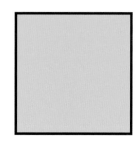

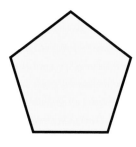

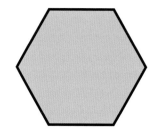

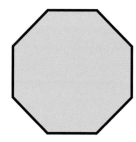

triangle =
3 angles and
3 sides

quadrilateral =
4 angles and
4 sides

pentagon =
5 angles and
5 sides

hexagon =
6 angles and
6 sides

octagon =
8 angles and
8 sides

Different types of triangle

The names of triangles are linked to their angles.

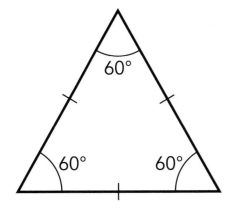 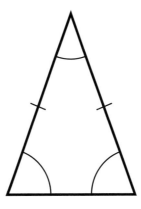 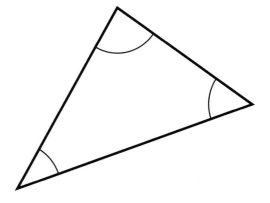

An equilateral triangle has 3 equal length sides and 3 equal angles (60°).

An isosceles triangle has 2 equal length sides and 2 equal angles.

A scalene triangle has 3 angles of different sizes and 3 sides of different lengths.

Isosceles triangles and scalene triangles can have right angles.

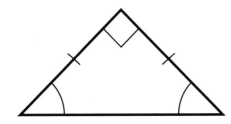 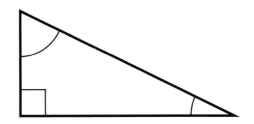

Symmetry

2-D shapes and images may have line symmetry. This is when the halves of the image are identical. It is sometimes called reflective symmetry because one half looks like a reflection in a mirror of the other half. That is why sometimes the line of symmetry is called a mirror line.

Images can have one or more lines of symmetry. They might have no lines of symmetry.

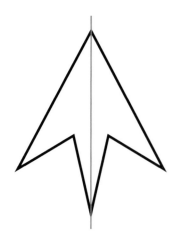

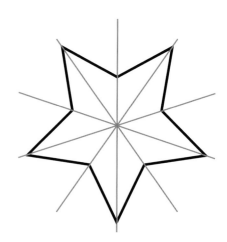

Equilateral triangles have
3 lines of symmetry.

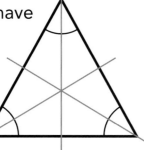

Isosceles triangles have
1 line of symmetry.

Area

Area is the amount of space covered by a 2-D object. It is the number of squares that the shape covers.

When this large rectangle is drawn on centimetre squared paper, it covers 7 rows of 10 squares – that is, 70 square centimetres.

The area of the small rectangle is 6 cm². So the area of the pink shape is 64 cm².

We can measure the area of rectangles, including squares, by multiplying the length by the width. The formula for this is $l \times w$.

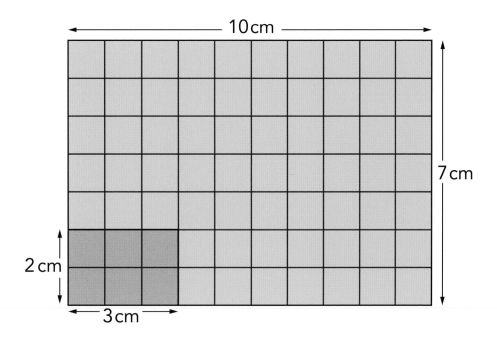

Kilometres and metres

We use kilometres to measure some lengths.

1000 m is equivalent to 1 km. Kilometres are useful units to use when lengths are very long.

If you run the length of a 100 m track 10 times, you will have run 1 kilometre.

A marathon is just over 42 km.

Perimeter

Perimeter is the distance around the outside of a shape. We measure perimeter using units of length, for example centimetres and metres.

The perimeter of this vegetable box can be found by adding together:

length + length + width + width.

We can find the perimeter of any regular shape by multiplying the length of one side by the number of sides that the shape has. For example, the perimeter of a square is 4 times the length of one side.

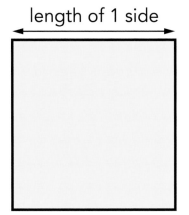

length of 1 side

perimeter = 4 times length of 1 side

If each side of this hexagonal swimming pool is 3 m, the perimeter is 3 m × 6, which is 18 m.

Perimeter of rectilinear shapes

Rectilinear shapes are shapes with sides that are perpendicular.
They are made up of straight lines and right angles.

These are rectilinear shapes:

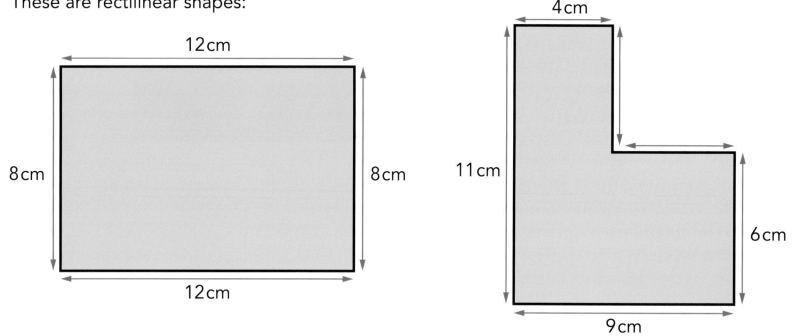

The perimeter of the rectangle is 40 cm.

We can find the perimeter of a rectilinear shape by adding the lengths of all its sides.
The perimeter of this rectilinear shape is 40cm.

If the shape is a rectangle, there are 2 pairs of equal sides so:

● we can use the formula 2*l* + 2*w* to find perimeter

● or we can add the width and length, then double that sum to find the perimeter.

Coordinates

To name a coordinate, find the horizontal axis first, then the vertical axis.

The blue square is at coordinate (D, 4).

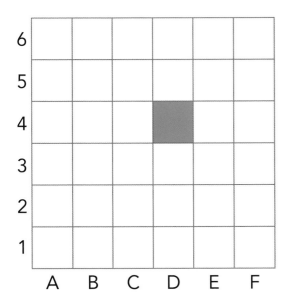

Time

We use different kinds of clocks.

This clock gives the time as 11 minutes past 10.

This clock shows 24 minutes past 9. We know it is a morning time because we can see the abbreviation a.m. If it was afternoon or evening the display would show p.m.

This is a 24-hour digital clock.

The time on this clock is 59 minutes past 11 in the evening. There is no need to show a.m. or p.m. because if the hour is between 12 and 24 we know it is between midday and midnight.

Money

| 1 penny | 2 pence | 5 pence | 10 pence | 20 pence | 50 pence | 1 pound | 2 pounds |
| 1p | 2p | 5p | 10p | 20p | 50p | £1 | £2 |

5 pounds
£5

10 pounds
£10

20 pounds
£20

50 pounds
£50

Calculating work rate

The most efficient packer in the factory has the fastest work rate.

- The work rate of each packer can be found by dividing the amount of work done by the time taken.
- Packer B is more efficient because she has a faster work rate. She packs 7 boxes per minute.
- Packer A is less efficient because he has a slower work rate. He packs 6 boxes per minute.

work rate = amount of work ÷ time taken

amount of work = work rate × time taken

time taken = amount of work ÷ work rate

33

The order of operations

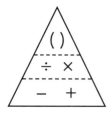

Remember, anything in brackets overrules the order of operations.

Laws of operations

The commutative law

$a + b = b + a$

$a \times b = b \times a$

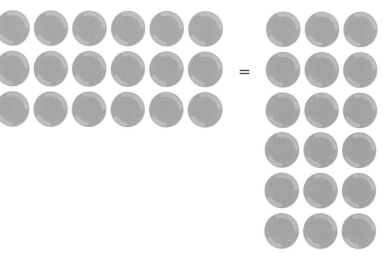

$$5 + 3 = 3 + 5 = 8$$

$$6 \times 3 = 3 \times 6 = 18$$

The associative law

$$a + (b + c) = (a + b) + c$$

$$a \times (b \times c) = (a \times b) \times c$$

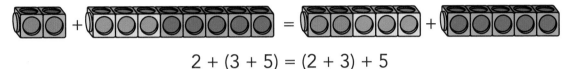

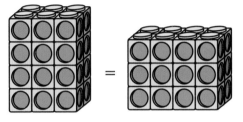

$$2 + (3 + 5) = (2 + 3) + 5$$

$$4 \times (2 \times 3) = (4 \times 2) \times 3$$

The distributive law

$$a \times (b + c) = (a \times b) + (a \times c)$$

$$a \times (b - c) = (a \times b) - (a \times c)$$

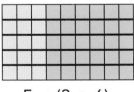

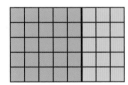

$$5 \times (3 + 6) \quad = (5 \times 3) + (5 \times 6)$$

$$5 \times (8 - 3) \quad = (5 \times 8) - (5 \times 3)$$

$$= 5 \times 5$$

12-hour time: time is displayed as 00:00a.m. to 11:59a.m. for midnight to noon, and 12:00p.m. to 11:59p.m. for noon to midnight

24-hour time: time is described using hours 0–24; midnight to noon is 00:00–12:00 and noon to midnight is 12:00 to 00:00

a.m.: Abbreviation that means 'before noon'. Used with 12-hour times to show times between midnight and noon.

addend: a number being added, or added to, in an addition calculation: addend + addend = sum

4 + 3 = 7
↑   ↑
addends

addition: An operation in which two or more numbers are combined or one number is increased by another. The symbol for addition is +.

43 + 16 = 59

analogue time: time shown using the hands on a clock

annual salary: money earned in one year

area: the amount of space covered by a two-dimensional object; the area of the green rectangle is 24 squares

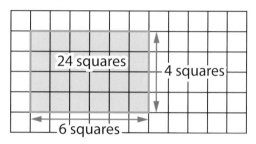

24 squares — 4 squares
— 6 squares —

array: an arrangement of items in rows and columns with equal numbers in each

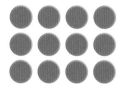

associative law: a rule in mathematics that states it does not matter which part of an addition or multiplication is completed first

$$a + (b + c) = (a + b) + c$$
$$a \times (b \times c) = (a \times b) \times c$$

bar chart: a graphical display of data in which the height of each bar relates to the number of items

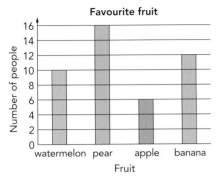

bar model: a diagram that represents known and unknown quantities as well as the relationship between these quantities as parts of a rectangular bar

9	
4	5

brackets: Any operation inside brackets in a number sentence overrules the normal order of operations. It is calculated first.

( )

centimetre (cm): a unit of measure equivalent to 10 millimetres

common year: a year that has 365 days; not a leap year

2019 CALENDAR

commutative law: a rule in mathematics which states that in addition and multiplication the numbers can be swapped around without changing the answer

$2 \times 50 = 100$ $\qquad$ $50 \times 2 = 100$

$a + b = b + a$ $\qquad$ $a \times b = b \times a$

continuous data: data that is measured; for example height, weight, temperature

data: information

decimal number: a number between two whole numbers

decimal places: the number of digits written in the decimal part of a number

decimal point: the symbol (a dot) used to separate the whole number part of a number from the decimal part

denominator: the number of equal parts an object, quantity or number has been divided into

difference: the result of a subtraction calculation: minuend − subtrahend = difference

digital time: time written in numbers

discrete data: data that is counted; for example the number of pupils in a school

distance: the length between two points

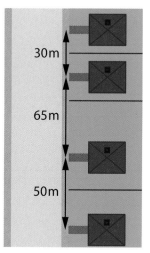

distributive law: A rule that means that adding a group of numbers together and then multiplying the total by something is the same as multiplying them separately and then adding them together. It is also true for division.

$$a \times (b + c) = a \times b + a \times c$$
$$\text{and}$$
$$a \times (b - c) = a \times b - a \times c$$

divide, division: split or share a whole into equal parts; the symbol for divide is ÷

$$\frac{3}{4}$$

dividend: the whole quantity before it is divided

In 540 ÷ 30 = 18, the dividend is 540

divisor: the number by which another number is divided

In 540 ÷ 30 = 18, the divisor is 30

double: twice as many, or twice as much

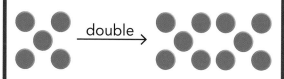

efficient strategy: a way of working out an answer quickly and accurately

equal: the same

equivalent: equal in quantity, size or value; have the same effect

estimating: judging an approximate result before calculating

factor: a number or quantity that, when multiplied with another, produces a given number

2 × 5 = 10

fraction: a number that is part of a whole

fraction wall: a visual representation that compares fractions of different sizes

gradient: how steep a line is

halve: separate into two equal parts

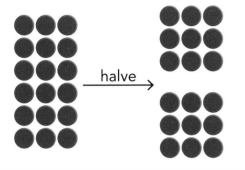

halve →

height: distance from base to top

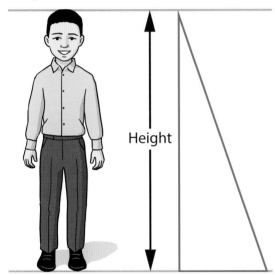

Height

horizontal axis: the reference line on a graph that runs from left to right through zero

hourly wage: the amount a person is paid per hour

Wages

10 hours at £7 per hour = £70

hundred: set of 100 ones

H	T	U
1	5	6

hundreds: In a three-digit number, the first digit shows the number of hundreds. A 'hundred' is a set of 100 'ones'.

hundredths: when 1 is split into 100 equal parts, each part is a hundredth

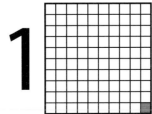

1 hundredth

inequality: a mathematical statement that identifies when two values are not equal

interval: The 'gap' between two numbers. This *y*-axis is labelled in intervals of 5.

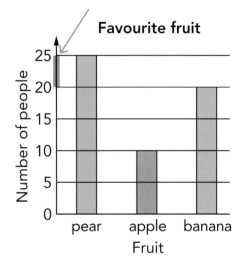

Favourite fruit

inverse: An operation that is the reverse or opposite of another operation. + and − are inverse operations. × and ÷ are inverse operations.

12 ÷ 3 = 4
is the inverse of
4 × 3 = 12

kilometre (km): unit of measurement of length equivalent to 1000 metres

Distance is measured in kilometres.

City ↑5km

leap year: A year that has 366 days; not a common year. Every 4th year is a leap year.

2020

January							February							March							April						

January
S M T W T F S
1 2 3 4
5 6 7 8 9 10 11
12 13 14 15 16 17 18
19 20 21 22 23 24 25
26 27 28 29 30 31

February
S M T W T F S
1
2 3 4 5 6 7 8
9 10 11 12 13 14 15
16 17 18 19 20 21 22
23 24 25 26 27 28 29

March
S M T W T F S
1 2 3 4 5 6 7
8 9 10 11 12 13 14
15 16 17 18 19 20 21
22 23 24 25 26 27 28
29 30 31

April
S M T W T F S
1 2 3 4
5 6 7 8 9 10 11
12 13 14 15 16 17 18
19 20 21 22 23 24 25
26 27 28 29 30

May
S M T W T F S
31   1 2
3 4 5 6 7 8 9
10 11 12 13 14 15 16
17 18 19 20 21 22 23
24 25 26 27 28 29 30

June
S M T W T F S
1 2 3 4 5 6
7 8 9 10 11 12 13
14 15 16 17 18 19 20
21 22 23 24 25 26 27
28 29 30

29

July
S M T W T F S
1 2 3 4
5 6 7 8 9 10 11
12 13 14 15 16 17 18
19 20 21 22 23 24 25
26 27 28 29 30 31

August
S M T W T F S
30 31   1
2 3 4 5 6 7 8
9 10 11 12 13 14 15
16 17 18 19 20 21 22
23 24 25 26 27 28 29

September
S M T W T F S
1 2 3 4 5
6 7 8 9 10 11 12
13 14 15 16 17 18 19
20 21 22 23 24 25 26
27 28 29 30

October
S M T W T F S
1 2 3
4 5 6 7 8 9 10
11 12 13 14 15 16 17
18 19 20 21 22 23 24
25 26 27 28 29 30 31

November
S M T W T F S
1 2 3 4 5 6 7
8 9 10 11 12 13 14
15 16 17 18 19 20 21
22 23 24 25 26 27 28
29 30

December
S M T W T F S
1 2 3 4 5
6 7 8 9 10 11 12
13 14 15 16 17 18 19
20 21 22 23 24 25 26
27 28 29 30 31

length: how long something is from end to end

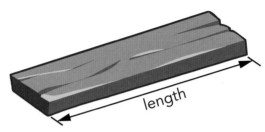

length

line graph: a graph in which connected data is plotted and connected by lines to show how something changes, often over time

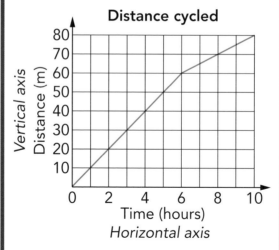

Distance cycled

Vertical axis — Distance (m)
80 70 60 50 40 30 20 10

0 2 4 6 8 10
Time (hours)
Horizontal axis

line model: a diagram that represents known and unknown quantities, as well as the relationship between these quantities, as sections of a line

20
Vans
Cars
How many vans and cars are there in total?

mental strategy: any method of working out the answer to a calculation in your head, without writing anything down

45 take away 6 equals 39 so I can use this to help!

450 – 60

metre (m): unit of length equivalent to 100 cm

1 metre

10 20 30 40 50 60 70 80 90 100

metre stick: an instrument used to measure 1 metre

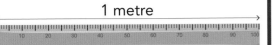

1 metre

millimetre (mm): unit of length; 10 mm is equivalent to 1 cm

minuend: the whole before parts are subtracted:

minuend − subtrahend = difference

minuend → (14) − 10 = 4

minute: unit of time equivalent to 60 seconds

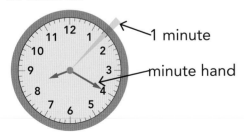

1 minute

minute hand

mixed decimal: a decimal number with a whole number part that is not zero

mixed operation: a calculation with more than one operation where the operations are not the same

multiple: a number that may be divided by another a certain number of times without a remainder

These are all multiples of 4.

$4 \times 1 = 4$
$4 \times 2 = 8$
$4 \times 3 = 12$
$4 \times 4 = 16$
$4 \times 5 = 20$
$4 \times 6 = 24$
$4 \times 7 = 28$
$4 \times 8 = 32$
$4 \times 9 = 36$
$4 \times 10 = 40$
$4 \times 11 = 44$
$4 \times 12 = 48$

multiplication: combining equal quantities

$$34 \times 3 = 102$$

multiplication grid: a table setting out multiples of all numbers, 1–10 or 1–12

1	2	3	4	5	6	7	8	9	10	11	12
2	4	6	8	10	12	14	16	18	20	22	24
3	6	9	12	15	18	21	24	27	30	33	36
4	8	12	16	20	24	28	32	36	40	44	48
5	10	15	20	25	30	35	40	45	50	55	60
6	12	18	24	30	36	42	48	54	60	66	72
7	14	21	28	35	42	49	56	63	70	77	84
8	16	24	32	40	48	56	64	72	80	88	96
9	18	27	36	45	54	63	72	81	90	99	108
10	20	30	40	50	60	70	80	90	100	110	120
11	22	33	44	55	66	77	88	99	110	121	132
12	24	36	48	60	72	84	96	108	120	132	144

multiply: repeat a quantity/value a number of times; the symbol for multiply is ×

number sentence: a mathematical sentence written with numbers and mathematical symbols

$$12 + 7 = 19$$

numerator: In a fraction, the numerator shows the number of parts. The numerators show how many parts – how many fifths, twelfths, quarters, eighths.

$$\frac{1}{5} \quad \frac{1}{12} \quad \frac{3}{4} \quad \frac{5}{8}$$

ones: in a 2-digit number, the second digit shows the number of ones

operation: a mathematical process; adding, subtracting, multiplying and dividing are all operations

+   −   ÷   ✗

order of operations: the order that operations should be completed in a calculation that has more than one step

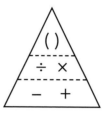

p.m.: Abbreviation that means 'after noon'. Used with 12-hour times to show a time between noon and midnight.

partition: splitting a number into parts that have the same total value

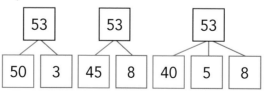

perimeter: distance around the outside of a shape

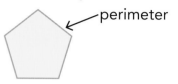

place value: The value of a digit is determined by its place. Place value can be thought of as 'quantity place value': knowing that the 3 in 35 is equal to 30; and 'column place value': knowing that the 3 in 35 is equal to 3 tens.

polyhedra: 3-D shapes; these have faces, edges and vertices

product: the answer when two or more numbers are multiplied together

$$45 \times 2 \times 2 = \textbf{180} \leftarrow \text{product}$$

pure decimal: a decimal number that has a whole number part that is zero

quotient: the result of dividing one number by another number

In $540 \div 30 = 18$, the quotient is 18.

rectilinear shape: shapes in which all sides are perpendicular

regrouping: re-partition tens and ones to help with calculating

24 has been regrouped as 10 and 14 so that 7 ones can be subtracted from 14 ones

Roman numeral: number written with symbols as they were in Roman times

IV, V, VI, IX, X, XI

rounding (to the nearest …): Writing a decimal number so that it has fewer digits. Numbers are rounded to the nearest approximate value.

ruler: an instrument used to measure lengths in millimetres and centimetres

scale: how the units are represented on a graph; the distance between marks

second: a very short period of time; there are 60 seconds in 1 minute

simplify: Laws and rules about operations can often be used to make a calculation easier. This is called simplifying.

$$54 \times 6 + 54 \times 4$$
$$= 54 \times (6 + 4)$$
$$= 54 \times 10$$

square centimetre (cm²): unit of measure of area

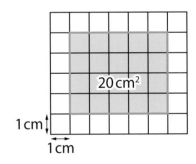

$20\,cm^2$

1cm

1cm

square metre (m²): unit of measure of area

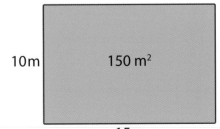

10m

$150\ m^2$

15m

statistical table: a table containing data

statistics: the collection and presention of data that can be used to answer questions

subtrahend: the number being subtracted from the minuend; minuend − subtrahend = difference

$$39 - 27 = 12$$

subtrahend

ten: set of 10 ones

tenths: when 1 is split into 10 equal parts, each part is a tenth

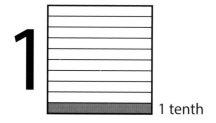

1 tenth

times: another word for multiply, 'lots of'

tree diagram: a diagram that looks like branches of a tree; used to represent calculations

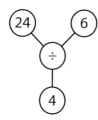

trend: a line on a graph that shows an upward tendency (something increasing) or a downward tendency (something decreasing)

unit fraction: a fraction in which the numerator is 1

unit price: what one item costs

$$90 \div 3 = 30$$

Unit price 30p

vertical axis: the reference line on a graph that runs from top to bottom through 0

width: how wide something is

work rate: The rate at which work is done. This is calculated by dividing the amount of work done by the time taken.

zero: The number before 1 in our number system. A number used to hold place value in a number.

405

In 405 there are no tens, but zero holds the tens place so we know that the digit 4 is worth 4 hundreds.

1	2	3	4	5	6	7	8	9	10
11	12	13	14	15	16	17	18	19	20
21	22	23	24	25	26	27	28	29	30
31	32	33	34	35	36	37	38	39	40
41	42	43	44	45	46	47	48	49	50
51	52	53	54	55	56	57	58	59	60
61	62	63	64	65	66	67	68	69	70
71	72	73	74	75	76	77	78	79	80
81	82	83	84	85	86	87	88	89	90
91	92	93	94	95	96	97	98	99	100

1	5	10	50	100
I	V	X	L	C

4	6	9	11	40	60	90
IV	VI	IX	XI	XL	LX	XC

18	34	59	92
XVIII	XXXIV	LIX	XCII